TROPICAL TREES

Found in
The Caribbean
South America
Central America
Mexico

•

Text and Color Photography

by

Dorothy and Bob Hargreaves

Published by
HARGREAVES COMPANY
BOX 895 • KAILUA, HAWAII 96734, U.S.A.

Price $2.00 in U. S. funds, postpaid anywhere in the world from the publisher

Also available at bookstores

AFRICAN TULIP TREE, Flame of the Forest, Fountain Tree

Spathodea campanulata BEAUVOIS

SP. Tulipán Africano, Llamarada de Bosque, Arbol de Fuente
FR. Immortel étranger, Baton de Sorcier
DUTCH. Tulpenboom

Discovered in 1787 by Palisot Beauvois in Africa, this ever-blooming fiery red tree can now be spotted throughout the tropical world. The name Fountain Tree comes from the water contained in the unopened buds (see picture at right), which squirt water when squeezed or pierced. Found in the Caribbean, South and Central America, and Mexico.

AKEE, Vegetable Brain

Sp. Seso Vegetal

Blighia sapida Koenig

This 50′ evergreen named after the English Mariner Captain Bligh of the "Bounty" and "Mutiny" fame, creates a gay Christmas-tree effect when the red fruits ripen. The 3″ to 4″ three-sectioned pods burst open when ripe. It is only when allowed to open naturally that the brain-like aril meat can be eaten. Attached to the tip of the three meat sections is the shiny black seed (see close up picture). The Akee is very popular as a food served in Jamaica and is delicious sautéed and served with codfish. Also seen in South America, St. Vincent Botanical Gardens, Hope Gardens, Fairchild Gardens, Florida, Dominica Botanical Gardens, Trinidad.

ALMOND TREE, Tropical Almond, Indian Almond, Malabar Almond

SP. Almendra, Almendro de la India, Alcornoque, Almendrón
FR. Amandier des Indes, Amandier Tropical, Zanmande
DUTCH. Manguel, Wilde Amandel, Amandelboom, Amanda
Terminalia catappa L.

As this 30′ tree withstands salt spray, it is seen extensively throughout the tropics in Southern Florida, the Caribbean, South and Central America, and Mexico. It can be recognized by its horizontal branches which grow in wide spreading circles at different levels on the trunk; by its large leathery leaves which turn red before they fall, suggestive of Autumn leaves; by its flat almond shaped edible fruits.

AMHERSTIA, Pride of Burma, Orchid Flower

Sp. Amherstia *Amherstia nobilis*

Said to be the finest flowering tree of the tropics, Amherstia is named after Lady Amherst, wife of a Governor of Burma. The tree was introduced into the botanical gardens in Trinidad in 1857. This tree (see picture at right) is one of the finest in the Western Tropics. The original tree was discovered by Dr. Wallich, Director of the Botanical Gardens, Calcutta, in the Teak forests of Burma. The beautiful vermilion colored blossoms which covered the ground were a daily offering to Buddha. The 8″ by 4″ flowers are superb. The many orchid-like flowers dangle from a central stem on each branch among the green foliage. Royal Botanical Gardens and Central Experimental Station (near airport) in Trinidad, Hope Gardens, Jamaica, Dominica Botanical Gardens.

ANNATTO, Arnatto, Lipstick Tree, Anatto

Sp. Achiote, Bija, Chaya, Cuajachote, Achote, Onoto
Fr. Roucou, Roucouyer, Achiot
Dutch. Rucu, Roucou, Koesoewee

Bixa orellana L.

Ancient warriors used this plant (Annatto Dye) to paint their bodies the orange red of the seeds. The Carib Indians in Dominica (Roucou), the Maya in Guatemala and Mexico, all used the dye extracted from the 50 seeds inside the soft spined fruits above. The dye has been used also to color cheese, margarine, candy, chocolate, cloth, lips, soap and paint. The pods also make interesting dried flower arrangements. The small evergreen tree has 2″ single pink flowers that resemble a wild rose. Grows throughout the Caribbean, South and Central America, and Mexico.

ANNONA FAMILY

SWEETSOP, Sugar Apple, Sweet Apple

Sp. Anón, Chirmoya, Saramuya
Fr. Cachiman Cannelle, Pomme Cannelle
Dutch. Scopappel, Kaneelappel

Annona squamosa L.

A 10′ to 20′ small deciduous tree, leaves 3″ to 5″, with a delicious fruit made up of numerous prominent knobs which, when ripe, can be pulled apart and the fruit and pulp eaten raw. Also made into drinks and sherbet.

SOURSOP

Sp. Guanábana, Guanaba, Catoche, Catuche
Fr. Corossolier, Corossol
Dutch. Sorsaka, Zuurzak, Soursap

Annona muricata L.

Another small 20′ evergreen West Indies Tree with fragrant, laurel-like leaves, and unusual kidney-shaped prickly fruits, weighing up to 6 pounds. A sweet-sour juice is extracted from the fruit's white pulp which makes pleasant cooling drinks and ices.

CUSTARD APPLE Bullock-heart

Sp. Corazón, Mamón, Anón, Anonillo, Anona Colorada, Anona de Redecilla, Chirimoya
Fr. Chachiman Coeur de Boeuf
Dutch. Kasjoema

Annona reticulata L.

This 25′ deciduous tree is distinguished from the other Annonas by its longer, narrow leaves 2″ by 8″ and smooth heart-shaped reddish-brown or yellowish 3″ to 5″ edible fruit with a network of hexagonal markings on its surface. All the Annonas have become naturalized throughout tropical America.

AUTOGRAPH TREE, Copey, Scotch Attorney, Clusia, Pitch Apple

SP. Cupey, Matapalo, Tampaco, Cape, Cucharo, Chuchi Copei
FR. Figuier Maudit, Figuier Marron, Aralie, Grande Feuille
DUTCH. Dam Machu, Cuchiú, Kopijk — *Clusia rosea* Jacq.

A West Indian evergreen tree with magnificent large thick ovate leaves 3″ x 8″ that withstand salt spray and high winds. These leaves can be used to autograph your name (see inset), the delight of lovers and children. It is said that the early Spanish conquistadores used them for playing cards and writing paper.

The wood is used mainly for fuel and construction, and medicine is obtained from the leaves, bark, and fruit. Seen throughout the Caribbean, Florida, Central and South America.

BANANA TREE

Sp. Guineo, Banano, Morado Plátano *Musa* (over 70 species)

There are over 300 edible forms of bananas. The trees are used for food, roofs, cattle feed, clothing, medicine, dye, alcohol, wine, vinegar, packing material, and many other things.

When the plant reaches flowering age, and the huge reddish bud emerges from its center, it extends downward, gradually opening up to expose tiered rows of tiny flowers. The upper ones become bananas, the lower ones furnish pollen, the middle ones drop off. Only one bunch is produced on each stalk. Once the plant has fruited it dies, but a new plant has sprung up around its base. This process takes about 18 months.

BANYAN—800 Species ficus (Fig)

BANYAN, Vada Tree, Indian Laurel

Sp. Higo, Hignera, Laurel de India *Ficus benghalensis* L.

This Indian Banyan is named for Hindu traders named Banyans. It is a sacred tree to the Hindus. The aerial roots (below) grow earthward from horizontal branches which support the huge evergreen tree that eventually covers large areas. One measured 2000 ft. x 85 ft. high in India.

BANYAN—Figs (Continued)

RUBBER TREE, Indian Rubber Fig, Rubber Plant, Assam Rubber Tree

SP. Palo de Goma, Goma Elástica, Higuera Caucho, Amate, Hule
FR. Caoutchouc *Ficus elastica* NOIS

A large spreading evergreen tree that reaches 100′ in India, its native home. It throws down thread-like roots (as do many of the ficus trees). The picture above shows the large, leathery, handsome leaves with the long pointed reddish sheath that covers each new leaf and soon falls. These new leaves are sometimes as large as 14″ x 7″. From the smooth grey bark comes the elastic substance called "Caoutchouc," which is one third the weight of the milky sap. In Assam, the original source of rubber, trees are first tapped when 25 years old, and yield until 100 years old. Now more valuable rubber is obtained in Brazil, called Para Rubber (Hevea). Many trees in El Salvador, Mexico, Puerto Rico, Virgin Islands, Martinique, Dutch West Indies and South America. Trees are used as ornamentals. Cuttings often used as house plants.

BAOBAB, Dead Rat Tree, Judas Bag, Monkey-Bread Tree, Bottle Tree

Sp. Baobab *Adansonia digitata* L.

From East Africa, this is one of the largest and longest lived trees in the world, ranking with the Sequoias. It has a thick trunk that grows up to 100′ around. Often, this trunk becomes hollow and can hold up to 250 gals. of water. It is one of Africa's most useful trees. The leaves, bark, and fruits are used for food, cordage, paper, thread, homes, medicine, and cloth. The 6″ pendulous white flowers and dark green foliage are handsome. The 12″ by 4″ long oval fruit hanging from a long stalk contains about 30 seeds suggesting the name "Judas Bag." Seen in Botanical Gardens in Florida, Dominica, St. Vincent, Trinidad and Tobago. There is a big tree in Queen's Park, Barbados.

BARRINGTONIA

SP. Tagua

Barringtonia asiatica (L.) KURZ
syn. *B. speciosa* FORST.

A tall handsome grey barked tree from Asia which is usually one of the first plants to obtain a footing on land, because the heart shaped fruits are buoyant and easily transported by the sea. The ovate 6″ to 18″ leaves form at the branch ends where the fragrant 5″ to 7″ brush-like blooms also cluster. They open in the evening, and the pink and white stamen blooms can be seen under the tree in the morning. The 4″ fruits contain seeds that can be grated and cast into the water to stun fish. Hope Gardens and Tower Isle Hotel Grounds in Jamaica, Botanical Gardens in Dominica, and Trinidad.

BAY RUM TREE, Bay Leaf Tree, Wild Cinnamon, Bayberry

SP. Ausú, Limoncillo, Malagueta, Ozúa
FR. Bois d'Inde, Français
DUTCH. Bay Boom

Pimenta racemosa syn. *P. acris*

Bay rum was originally prepared by distilling the leaves of this 40′ evergreen tree in rum. The main product is Myrcia Oil (Bay Oil) obtained by distillation of the leaves and twigs to make bay rum for cosmteics and medicine (1 lb. of oil from 100 lbs. of leaves). A large forest produces oil on St. John in the Virgin Islands. Trees also in S. Florida, Bermuda, Bahamas, Jamaica, Grenada, Trinidad, Tobago and South America.

BEAD TREE, Jumbie-Bead, Sandal Bead, Red Sandalwood, Circassian

SP. Peronías, Sandalo, Coralitos, Coral, Palo de Mato, Coquelicot
FR. Réglisse, Arbre à Réglisse, Arbre à Graines Réglisse, Corail Végétal

Adenanthera pavonina L.

A 40′ tall erect tree from Ceylon has branches of fine feathery foliage high on the smooth grey trunk. The wood is strong and durable, called red sandalwood because it is sometimes substituted for real sandalwood. It has small yellow flowers and curling pods which can be used for dried flower arrangements, while the bright red round seeds they contain can be softened in boiling water and used for necklaces and other novelties (see inset). They are called Circassian or Barricari seeds and are eaten and used as weights in Asia.

BREADFRUIT

SP. Arbol de Pan, Palo de Pan, Fruita de Pan, Mazapán, Panapén
FR. Arbre Végétal, Fruit à Pain
DUTCH. Broodboom, Palu di frut'i pan

Artocarpus incisus syn.
A. altilis, A. communis.

The breadfruit was introduced into the West Indies in 1793 from Tahiti. Captain Bligh in the ship *Providence*, brought plants to St. Vincent and Jamaica after an earlier attempt failed aboard the mutiny ship *Bounty* in 1789. This 60′ evergreen tree has handsome looking deep lobed shiny dark green leaves from 1 to 3 feet long. The seedless variety of fruit 5″ to 8″ weighing up to about 10 lbs. has a tough yellowish green warty hide—smoother than the seeded variety (see next page). Fried, roasted, or boiled, the fruit is high in carbohydrates and a source of Vitamins A, B, and C.

BREADNUT TREE

SP. Pana de Pepitas, Castaña, Castaño del Malabar, Chataigne
FR. Châtaignier, Châtaignier du Pays

This variety of *Artocarpus*, a breadfruit distinguished by its large brown edible seeds and rougher surface than seedless variety, is composed of greenish conical spinelike projections. Its leaves are less deeply lobed and are more hairy. Flowers and fruits are borne throughout the year. Called Bread Nut of Dominica. (Not to be confused with the Bread Nuts of Jamaica—*Brosimum*, a 1″ nut.)

BUTTERCUP TREE, Wild Cotton, Yellowsilk, Shellseed, Brazilian Rose

SP. Rosa Imperial, Rosa China, Emperatriz de la Selva, Tecomasuche, Apomo Panaco, Pochote, Pumpo
DUTCH. Njoe Fodoe

Cochlospermum vitifolium (Willd.) SPRENG

Outstanding because of its bright yellow blossoms that resemble large buttercups or roses, this ornamental is cultivated from Florida, Puerto Rico, the Virgin Islands, through the West Indies, Central America, South America and Mexico.

After the five 3″ to 5″ golden cups fade, the 3″ fruit capsule forms. This contains a number of kidney-shaped seeds covered with long white floss, the "Cotton or Silk" of the tree. This is used like kapok to stuff pillows and mattresses. The tree is deciduous and usually blooms when leafless. The grapevine-like leaf of the tree gives it the Latin name viti (grape) folium (leaf). In Puerto Rico there is a double flowered petal variety which usually blooms in February and March. Called Carnestolendas in Venezuela (probably because it blooms during Lent).

CALABASH TREE, Gourd Tree

SP. Higüero, Güira, Calabasa, Cujete, Guaje, Morro, Guacal, Totumo, Calabazo
FR. Calebasse, Calebassier
DUTCH. Kalebas, Kalabassenboom, Calbás Rondo *Crescentia cujete* L.

There are many of these unusual 30′ evergreen trees in Jamaica, Puerto Rico, Guatemala, Martinique, Dominica, the Dominican Republic and other tropical American countries. They can be identified by their short trunk with long spreading branches, leaves that seem to jut out from these branches, unpleasant smelling 2″ x 3″ yellow and purple flowers borne on the trunk and branches, and by the numerous roundish green fruits 5″ x 12″ (note on tree below) which first turn yellow, then brown. These hard shells can be beautifully polished as the gourd and maraca at right. They can be trained into different shapes by tying them when green.

CAMPHOR TREE

SP. Alcanfor

Cinnamomum camphora

This 40′ native of China and Japan furnishes the well known camphor of commerce obtained by distilling the wood, roots, twigs, and leaves.

Some trees are oil bearing types, and some yield the solid camphor. The tree pictured is from the Botanical Gardens in Dominica, and is the oil bearing type. Both trees found also in Trinidad, St. Vincent, Puerto Rico. The shiny, dark green leaves form a dense crown for this tall tree, and when crushed have the odor of camphor. Linen stored in containers of camphor wood keeps well. It is believed that this odor helps keep things insect proof.

In recent years the production of camphor from turpentine has affected the market for the natural camphor.

CANDLE TREE

SP. Arbol de la Vela, Xkat-cunc, Kaat, Pepina
Parmentiera cereifera SEEM

A tree from Panama that branches close to the ground and has curious 1′ to 3′ long candle-like fruits. They have a waxen texture and resemble hand dipped candles. They also have a pungent, apple-like fragrance. The greenish-white flowers develop on the main trunk and larger branches the same as those of the Calabash Tree. Seen in Botanical Gardens in Trinidad, Dominica, St. Vincent, University in Venezuela.

CANNONBALL TREE

Couroupita guianensis AUBLET

SP. Bala de Cañón, Coco de Mono, Granadillo, Cuirana, Moke
FR. Arbre a Bombes, Boulet de Canon, Abricot de Singe
DUTCH. Boesi, Kalabasi, Boseh Kalabas, Koppe Jewadaballi, Boskelebas

This huge tree is a novelty wherever it grows. The 3″ to 5″ waxy sweet smelling flowers seen in the picture below, push right out of the heavy bark and have no connection with the foliage at the top of the tree. These are followed by hard shelled cannonball like fruits 6″ to 8″ (used for calabashes); the pulp has an unpleasant odor. The long dangling fruiting branches give the tree an unkempt appearance. Seen in Botanical Gardens in Florida, Dominica, Trinidad, St. Vincent, British Guiana, and throughout Central, South America and Mexico.

CASHEW NUT, Cashew Apple

SP. Marañón, Cajuil, Caju, Jocote Marañón, Merey, Pajuil
FR. Pomme Cajou Acajou, Noix d'acajou, Pommier d'acajou
DUTCH. Cashu, Kasjoe, Palu di Cashupete, Orvi, Boschkasjoe
Anacardium occidentale L.

This tall spreading evergreen thrives in open areas. It has tiny fragrant pink flowers (see center of picture) that develop a kidney-shaped seed, edible only when roasted. This is attached to a reddish pear shaped fruit (see top of picture) which is also edible. The hulls of the nut contain an oil, Cardol Oil, which blisters the skin. Grows throughout the tropical world.

CACAO, Cocoa, Chocolate Tree

SP. Cacáo, Caceteiro, Caca-uatl, Truijillano
FR. Cacao, Cacaoyer

Theobroma cacao L.

Cacao is one of the best known plants originating in the New World. The Spanish conquistadores found it cultivated by the Mexican Indians. Grown extensively in wet tropical regions of Dominica, Trinidad, and other islands of the West Indies, as well as Central and South America, and Mexico. The large red pods hang from the small 25′ evergreen tree, shaded by protective trees (see Poui Trees). These large five-celled fruits contain many 1″ bitter tasting seeds or beans from which the cocoa and chocolate is made by roasting and grinding. (200 seeds to a pound.)

CASSIA, Pink Shower Cassia, Coral Shower Tree

SP. Canafistula Cimarróna
FR. Casse

Cassia grandis L.f.

There are many Cassias that bloom throughout the tropics. (See "Tropical Blossoms of the Caribbean"—*Hargreaves*.) However, this coral colored Cassia is the first to bloom in the spring. The striking coral-pink blossoms massed in clusters along its branches make the tree a beautiful sight to see, but they soon fall, forming a colorful carpet on the ground below. The tree has large black 3′ pods that contain flat round yellow seeds embedded in dark, sticky, bitter, unpleasant smelling pulp with laxative properties. The 50′ tree has smooth grey bark and widespreading branches. Fairchild Gardens, Florida, Dominica Botanical Gardens, and many other tropical areas.

CASUARINA, Australian Pine, Willow, Beef Wood, Whistling Pine, Horsetail

SP. Pino, Pino Australiano, Ciprés
FR. Pin d'Australie, Filao
DUTCH. Casuarine

Casuarina equisetifolia L.

A leafless 100′ evergreen, not really a pine, named for a large Australian bird called a Cassowary. The long, drooping dull green needles are tufted like the feathers of the bird. Some trees growing in Malaya are said to be several hundred years old.

The tiny red tufted flowers and cones, seen below, appear in May and June. These cones are used for making novelties. The trees are useful as windbreaks; their sturdiness a symbol of faithfulness. They can be trained into interesting shapes and hedges, or can be used for reforestation. Grown from Florida, Bermuda, Bahamas, throughout the West Indies, Trinidad, Aruba, South America and Mexico.

CEIBA, Kapok, Silk Cotton Tree, Cotton Tree

SP. Ceiba, Ceibo, Pochote, Bongo, Ceiba de Lana
FR. Mapou, Fromager, Bois Coton, Kapokier
DUTCH. Katoenboom, Katunbom, Kapokboom, Kankantri

Ceiba pentandra (L.) GAERTN.

One of the largest trees in Tropical America is the National Tree of Guatemala, held sacred by the Ancient Mayas. "History was said to have its beginning with the Sacred Ceiba Tree," from which, according to Mexican Maya legend, "the first man was born." The huge thick trunked trees, with branches sticking out at right angles, are often planted in the market square for their majestic shade. The leaves have 5 to 9 fingers, and the creamy white 1″ flowers appear just before the leaves. Two months after they fall, black oblong seeds 3″ to 6″ form. These capsules contain the floss called kapok used to stuff mattresses, pillows, life preservers, and upholstery.

CHICLE, Chewing Gum Tree, Sapodilla, Naseberry, Dilly

SP. Níspero, Zapote, Chicozapote, Zapotillo, Muyozapot
FR. Sapotillier, Sapotille
DUTCH. Mispu, Mispel, Sapotille, Mispelboom, Sapatija

Manikara zapota syn. *Achras zapota*

This handsome evergreen about 60′ high, is noted particularly as a source of chicle for chewing gum. The fruit is a favorite in Tropical America. Besides being delicious raw, it can be made into preserves and syrup. The flesh is sweet tasting inside the roundish brown 3″ rough skinned fruit.

Gum is obtained by tapping the trees every two or three years. Yields about 60 quarts of latex. Mexico, British Honduras, Guatemala, South America, Central America, West Indies.

CHINABERRY, Pride of India, Indian Lilac, Bead Tree

SP. Lilaila, Pasilla, Paraíso, Lila, Lilayo, Violeta, Jacinto, Alelí, Alilaila
FR. Lilas, Lilas du Pays, Lilas des Indes
DUTCH. Alelí, Anesita

Melia azedarach L.

Legend has it that the Chinaberry Tree has been worshipped in the Temples of Persia, Malaya, and Ceylon for as long as man can remember, yet, in some places in the world, it is considered a weed.

Fast growing to 40′, it blooms in the spring with giant clusters of fragrant flowers. Each pink or blue lilac-like blossom has a tiny dark purple staminal tube sticking out of its middle. The tree produces abundant fruits about the size of a cherry with a single stone which can be used as a bead, giving the tree the name Bead Tree. The tree grows rapidly, sprouting easily from stumps, cuttings, seeds, but is short lived, and the brittle limbs are easily broken by the wind. The wood is long lasting, so is used for musical instruments. Other parts of the tree have medicinal properties. Tropical countries, from S. Florida to South America.

CHORISIA, Floss Silk Tree

SP. Ceiba del Brasil

Chorisia speciosa

When in bloom in the Autumn, this unusual, large tree is lovely to see, but, when not in bloom, the smooth, stocky green trunk is usually covered with stubby spines, and its sprawling angular branches aren't so pretty. The striking 3″, five petaled blossoms are crimson, or pink, or white. According to Edwin A. Menninger, "The flowers of no two of these trees are exactly alike." Pear shaped capsules follow these blooms, and when they burst open, they release the silky white floss that gives the tree its name. This also is used for kapok. Fairchild Gardens, Florida, Cuba.

COFFEE TREE, Arabian Coffee

SP. Café, Cafeto
FR. Caféier, Café
DUTCH. Koffie

Coffea arabica L.

Coffee is one of the most important products exported from Tropical America. It is said that Coffee was introduced in Surinam in 1714, Jamaica in 1718, Martinique in 1720, and Brazil in 1727.

The Coffee Tree is a small evergreen 12 to 15 feet high with lovely shiny dark green leaves 3″ to 6″, and flowers that are very fragrant (related to the Gardenia).

The berries ripen in the fall and are harvested, about 1000 seeds to a pound. The thin fleshy pulp is then removed, the beans dried, shelled, graded, and stored to season, then roasted and ground. Grown in Bermuda, Puerto Rico, West Indies, Central and South America, and Mexico.

COLVILLEA, Colville's Glory

SP. Colvillea
Colvillea racemosa BOJER

The foliage of this tree resembles the Poinciana, a relative, but the fern-like leaves of the Colvillea are larger. Both trees are from Madagascar. The rich orange blossoms burst into bloom in the fall when there aren't too many other trees blooming. They grow in large clusters at the branch tips. The buds resemble huge orange clusters of grapes. These buds gradually pop open revealing the true flower with tiny orange stamens (see top of picture above). Seen in Fairchild Tropical Gardens, Florida, Botanical Gardens in St. Vincent, Dominica, Trinidad and Tobago.

CORDIA, Geiger Tree, Anaconda, Geranium Tree

SP. Vomitel, Cutiperi, San Bartolemé, Capá
FR. Bois de Rose, Bois de Rhodes, Chêne Caparo
DUTCH. Carawara Español

Cordia sebestena L.

As this evergreen tree from Tropical America flowers almost continuously, and also withstands salt air, it is used widely for ornamental purposes. It requires little water so thrives on some of the more arid islands. There are many beautiful large trees on Barbados, Antigua, Curacao, Aruba, and other tropical areas from Bermuda, the Bahamas, Puerto Rico through South America, and Mexico.

CREPE MYRTLE, Crape Myrtle, Queen of Flowers, Queen Flower, June Rose, Pride of India

SP. Astromelia, Reina de las Flores, Astromero

Lagerstroemia speciosa (L.) Pers.

Speciosa comes from the Latin meaning pleasing to the eye. This beautiful deciduous tree, native of India, is certainly pleasing to the eye. It is valued for its tough red timber, medicinal use, and ornamental beauty. Many in S. Florida, Jamaica, and other Caribbean Islands, Puerto Rico, Trinidad, Central and South America, and Mexico.

DIVI-DIVI, Livi-Divi

Caesalpinia coriaria

A small tropical American evergreen of about 30′ with feathery foliage and fragrant flowers. Picture was taken on Aruba where this interesting windblown tree has become a symbol.

The small curled pods (see close up), contain tannin which formerly was exported. The tiny yellow flowers attract numerous bees. Besides the many trees on Aruba, the Divi-Divi can be seen also in the Botanical Gardens in Dominica, Trinidad, Florida, also in the Virgin Islands, Antigua, and in South America.

DOMBEYA, Hydrangea Tree, Pink Ball, Mexican Rose

SP. Bedonia

Dombeya wallichii

Showy, drooping, pink, scentless balls, like Christmas ornaments, bloom at Christmas time among the dense foliage of this striking tree. Each 1″ flower has five wide petals. Many of these flowers are crowded into the showy heads that form these lovely clusters. They continue to hang on their long hairy stalks long after the color has faded and they have turned to brown. Found at the Experimental Station near Tegucigalpa in Honduras, Botanical Gardens in Guatemala City, Fairchild Gardens, Florida.

ELEPHANT'S EAR, Earpod, Mulatto Ear, Monkey Ear

SP. Guanacaste, Oreja de Mulato, Oreja de Mono, Oreja de Vaina

Enterolobium cyclocarpum (Jacq.) Griseb.

A Venezuelan spreading deciduous tree of the Mimosa Family with a huge grey 10′ trunk and branches. In the dry season, when bare of its feathery foliage, the curious brown ear-shaped 3″ seedpods can be seen dangling high on the tree. Seeds can be used as necklaces, pods are relished by cattle, and young pods and seeds can be used for human food. The wood is used for cabinets, boats, and the bark is used for tannin, soap, and medicine. Found in Mexico, Virgin Islands, Guatemala, Honduras, Dominica, and many of other Caribbean Islands, Central and South America.

ELDER (Yellow), Yellow Trumpet, Yellow Cedar, Ginger Thomas

SP. Saúco Amarillo, Roble Amarillo, Ruibarba, San Andrés, Tronadora, Retama, Gloria, Copete, Fresnillo
FR. Chevalier, Bois Fleurs Jaunes, Bois Caraibe
DUTCH. Kelki Heel, Gelki Heel (Yellow Blossom) *Tecoma stans* (L.) H.B.K.

This bright yellow small evergreen 30′ tree often blooms along the roadsides in the Bahamas, Caribbean Islands, Central and South America and Mexico, sometimes twice a year. It is the official flower of the U.S. Virgin Islands and also the Bahamas.

EUCALYPTUS, Gum Tree, Beakpod, Swamp Mahogany, Browngum

SP. Eucalipto de Pantano, Eucalipto del Alcanfor
(There are about 300 species of Eucalyptus)

Eucalyptus robusta J.E.Sm.

A handsome native of the Myrtle Family from Australia with thick alternate pointed leaves and tiny cup-like fruiting capsules (see picture at left). Uses include timber, gum, tannin oil for medicine (eucalytpus oil comes from the leaves). Grows from Puerto Rico, Trinidad, Mexico to South America.

FLAMBOYANT, Royal Poinciana, Flame Tree, Flame of the Forest

SP. Flamboyán, Flamboyán Rojo, Arbol de Fuego, Tabuchín, Guacamaya
FR. AND DUTCH. Flamboyant.

Delonix regia (Bojer) Raf.

One of the most strikingly beautiful trees of the world, it is known by the same name throughout most of the world. It is the National Flower of Puerto Rico.

Like a huge scarlet umbrella, the tree bursts into dense clusters of scarlet or orange blossoms early in the spring that last until late summer. One petal of the five petals of each bloom is white. Soon the lovely fern-like foliage appears followed by long brown pods whose seeds can be used in necklaces.

GUAVA

SP. Guayaba, Guayava, Guayabo, Guayaba Perulera *Psidium guajava* L.
FR. Goyave, Goyavier, Goyevier à Fruits
DUTCH. Goejaba, Guave, Goeajaaba, Guyaba

This small 25′ evergreen is commonly cultivated (often wild) as a fruit tree throughout the tropics because of the delicious jam, pastes, jellies and juice made from its fruits. They are rich in Vitamin C, iron and calcium. The tree has prominently veined 3″ to 6″ oblong leaves, fragrant 1″ white flowers, followed by the lemon-like thin skinned aromatic fruits. Pulp, which surrounds a mass of small hard seeds, is white or pink.

JACARANDA, Fern Tree

SP. Jacarandá, Flamboyán Azul, Palo de Buba, Guarupa
FR. Flamboyant Bleu, Barbatimao, Samarapa

Jacaranda acutifolia syn. *J. mimosifolia*

The cool blue blossoms of this large Bignonia Tree from Brazil are a welcome sight in the tropics. It is one of the few trees commonly called by its botanical name throughout the world. The lovely fern-like leaves form a ceiling above the handsome grey barked trunk and branches, and when the little blue bell-like flowers fall, they form a carpet under the tree. Grows profusely in Mexico, Central America, South America. Some also in Florida, Puerto Rico, the Virgin Islands, and the Caribbean.

JACK FRUIT TREE, Jac Fruit

SP. Jaca, Pana Cimarrona, Pan de Fruta, Buen Pan, Albopán, Castaño, Rima, Jaqueira.
FR. Jaquier

Artocarpus heterophyllus Lam.

This strange relative of the Breadfruit comes from India and Malaysia. The tree grows to a height of fifty feet. The fruits are most unusual. They weigh up to seventy pounds each, are up to three feet long, and are borne all along the trunk of the tree.

These fruits are an important food in the Eastern Tropics. Although the ripe fruit has an unpleasant odor, the taste outweighs this drawback. The yellowish soft flaky, sweet pulp is eaten raw boiled, or fried, and is delicious in curries. The large white seeds are also good roasted, tasting something like chestnuts. Found throughout the Caribbean, Central and South America, and Florida.

LIGNUM VITAE, Tree of Life

SP. Guayacán, Palo Santo, Guayacán de Playa, Guayaco
FR. Gaïac, Gaïac Franc, Gaïac Mâle, Bois Saint
DUTCH. Wayacá, Pokhout
Guaiacum officinale L.

Sixteen years after Columbus made his first voyage, the Spanish established this small evergreen tree as an article for trade. It has been a source of resin for centuries. Lignum Vitae means "wood of life." It is the heaviest of all woods and will sink in water. Useful for woods used under water. The dense crowned slow growing 30′ tree has clusters of small felty blue flowers at the branch tips, and orange heart shaped fruits follow or appear at the same time. There are many of these trees in Jamaica where it is their National Flower. Also some in Dominica, St. Vincent, Trinidad, Tobago, Central and South America, Mexico, Curacao and Aruba.

MAHOGANY, West Indies or Dominican Mahogany

SP. Caoba, Caoba de Santo Domingo
FR. Acajou, Mahogany Petites Feuilles
DUTCH. Mahok, Mahoni
Swietenia mahagoni (L.) Jacq.

The oldest surviving use of Mahogany is said to be in the Cathedral at Santo Domingo, 1550. The Mahogany flower is the Dominican Republic's National Emblem. Native of the West Indies, this tall 75′ evergreen with lovely hard wood, is valued for furniture, finishing houses, ships, and cabinets. Honduras Mahogany *(S. macrophylla)* is now more important commercially. It has larger leaves and fruit. Both have large pear shaped seed capsules.

MALAY APPLE, Mountain Apple, Ohia, Otaheite-apple, Pomerac

SP. Pomarrosa, Manzana, Pomagás, Pomagada, Pomme Malac, Marañón Japonés, Pomarrosa de Malaca, Pera
FR. Jamelac, Pomme de Tahiti, Pomme de Malaisie
DUTCH. Pommerak, Djamboe Bol

Eugenia malaccensis L.

Malay Apple was introduced into Jamaica in 1793 when Captain Bligh, aboard the *Providence*, was commissioned to bring in fruit from Tahiti. According to Polynesian legend, the tree was sacred. Temple idols were carved from its wood. (See "Tropical Trees of Hawaii"—*Hargreaves.*) The Native Malayan tree is a lovely 50′ evergreen with handsome large shiny green leaves 7″ to 12″ long, and tiny cerise blossoms that look like wee shaving brushes. These pop out all over the trunk and limbs of the tree. Soon the little pear-shaped 2″ to 3″ pink and white apple-tasting fruit develops. This can be eaten raw or cooked for preserves or wine. Widely planted throughout the tropics.

MAMMEE APPLE, Mamey, Apricot Mamey

SP. Mamey, Mamey Amarillo, Zapote de Niño, Mamey de Santo Domingo, Ruri, Mamey de Cartagena, Abricoteiro
FR. Abricotier, Abricot, Abricot des Antilles, Abricot de Saint-Dominque, Abricotier d'Amérique, Abricot Pays
DUTCH. Mami, Mamaya, Mammi, Mamieboom, Mamaja

Mammea americana L.

Another stately 60′ West Indies evergreen of the Mangosteen Family, with 4″ to 8″ shiny leathery dark green leaves, 1″ white fragrant flowers, is best known for its 3″ to 8″ spherical thick skinned brown edible fruits. Their pulp is orangish red, sweet and can be eaten raw or cooked. Each "apple" has 2 to 4 very large oblong reddish-brown seeds.

The flowers also are used in the French West Indies for a liqueur known as "Eau de Créole" or "Créme de Créole."

The juice and seeds are used in Mexico as insecticides. *Calocarpum sapota* is often wrongly called Mammee. Cultivated from Bermuda, S. Florida through the Bahamas, West Indies, Trinidad, South and Central America, and Mexico.

MANGO

SP. Mangó, Mango, Manga
FR. Mangot, Mangotine, Mangue, Manguier
DUTCH. Manja, Kajanna Manja, Bobbie Manja, Manggo, Mangga-boom

Mangifera indica L.

Originating in India, the Mango is now planted throughout the tropics. It is known as the "king of fruits." An evergreen, it is widely planted as a shade and fruit tree, growing to a dense large 65′ tree (see center picture—note Water Buffalo in foreground).

It flowers usually in winter to spring when tiny pinkish hairy flowers form at the branch tips (see below). Some people are highly allergic to the blooms and skin of the Mango. The maturing fruits (above) hang prolifically from the tree March to October. Their orange pulp is deliciously sweet-tasting, juicy, and is one of the most popular fruits of the tropics. It is said that Mangos have been cultivated by man for 4000 years.

Besides being eaten raw, they can be stewed, frozen, and made into very fine chutneys. They are thought to taste like a peach, apricot, cantaloupe, or even a pineapple.

The leaves smell like turpentine when crushed, and have even been used in Mexico to clean teeth and harden the gums.

NAPOLEON'S HAT, Napoleon's Button

SP. Napoleona

Napoleona imperialis

A curious flower growing on a small evergreen African tree was dedicated by a Frenchman named Beauvois to Napoleon Bonaparte in 1808. The remarkable flower, somewhat resembling the Passion Flower, has a double crown, first composed of 35 to 40 yellow thongs, while the second is a more rigid crown; its yellow points rimmed with purple. These sit in an interesting five-lobed reddish-purple cup. The smooth leathery 6″ x 2″ leaves are something like Camellia leaves. The tree pictured here is in the Royal Botanical Gardens in Trinidad. It has produced seeds only once, in 1923.

NORFOLK ISLAND PINE

SP. Pino "Norfolk," Eucaria
FR. Pin d'Norfolk

Araucaria excelsa

This is a perfectly symmetrical evergreen tree discovered by Captain Cook on Norfolk Island near Australia in the Pacific Ocean. Actually, the trees are not pines. They do not have needles like a pine, but overlapping scale - like leaves about one half inch long. These trees, up to 200′ tall, are sometimes used for masts on ships. They are used in landscaping and as Christmas trees throughout S. Florida, the Caribbean, and South America. There are many in Panama, Guatemala, El Salvador and other Central American countries also. A closely related *Araucaria* is the Cook Pine, *A. columnaris*, which comes from New Caledonia (Isle of Pines). Captain Cook, who also discovered this tree, said upon approaching the island, "They had the appearance of tall pines which occasioned my giving that name to the island." It is hard to distinguish these two trees apart.

NUTMEG, Sadhika

SP. Nuez Moscada

Myristica fragrans Houtt.

Native of the Moluccas (Spice Islands). A large evergreen tree 70′ to 80′ tall that likes humid conditions. The small, pale yellow flowers have no petals, but they produce the fruit which has a kernel that is so valued as the nutmeg spice. The ovid, yellow 2″ fruit, resembling an apricot, about 5 or 6 months after flowering, ripens and splits open disclosing the glossy, dark brown seed, the nutmeg kernel, inside. This is surrounded by a net-like scarlet aril which is "Mace," another valued spice. (See close up above.) 100 nutmegs produce about 3 oz. of dried mace. Found in Trinidad, Grenada, Dominica, and St. Vincent Botanical Gardens, and there are some large trees in Welchman Hall Gully in Barbados.

ORCHID TREE, Bull-hoof, Mountain Ebony, Poor Man's Orchid, Napoleon's Hat

SP. Palo de Orquídeas, Urape
FR. Deux Jumelles, Caractére des Hommes
DUTCH. Vlinderbloem

Bauhinia variegata L.

Named after the Bauhins, who were two medical brothers dedicated to botany, rather than medicine. Grows readily in all tropical countries. It was brought to the Caribbean in the 1600's. It is a small deciduous tree with variegated lavender orchid-like blooms and cleft leaves that resemble the hoofs of animals. Has a 2 ft. long smooth brown pod.

PALM TREES
COCONUT TREE, Coco Palm, Coconut Palm

SP. Palma de Coco, Cocotero, Palmera de Coco, Coco de Aqua
FR. Coco, Noix de Coc, Cocotier
DUTCH. Klapperboom, Coco, Cocos, Cocospalm

Cocos nucifera L.

There are 1500 species of Palms, so only a few are noted. The Coco Palm is the best known, and one of the most beautiful Palms in the world, its origin dating back to prehistoric times. They retain their germinating power 4 months after floating in the ocean. Growing to 100′, bearing over 40 fruits yearly at 6 years, they furnish food, oil for soap, cosmetics, margarine, Vitamin B, decorations, mats, hats, shelters, shade, novelties. Rank among ten most useful trees in the world.

Palm Trees (cont.)

ROYAL PALM, Puerto Rico Royal Palm, Mountain Cabbage

SP. Palma Real, Cubana, Palma de Costa, Palma de Yaguas

Roystonea elata syn. *R. regia*

A very stately native of Cuba named for Gen. Roy Stone, an American engineer in Puerto Rico. The trees are smooth ringed, light grey, with straight, majestic 60′ trunks, often planted in rows.

QUEEN PALM, Feather Palm, Feathery Coconut Palm, Monkey Nut

SP. Coco de Pluma

Arecastrum romanzoffianum syn. *Cocos plumosa*

Feathery plumes bedeck this 30′ to 60′ ornamental from South America. The 3′ to 6′ flowering clusters produce fruit with sweet tasting orange pulp. These, when peeled, have the same little monkey face the coconuts have. Can be used for jewelry (see page 61).

DATE PALM

SP. Dátiles

Phoenix dactylifera L.

Prehistoric. Date crop one of the oldest known—5000 years. 800 uses, hats, mats, thatch, food, wine sugar, etc. Bears from 5 years until 200 years. Requires little water. Bears 10,000 flowers in each cluster, 300 lbs. of dates to a tree. High in food value—50% sugar, 7% protein. Mohammed said to the Arabs: "Honor the Date Palm for it is your mother." Christendom's symbol of martyrdom or victory.

PAPAYA, Pawpaw, Papay, Papaw

SP. Lechosa, Papaya, Fruta Bomba, Melón Zapote
FR. Papaye, Papayer
DUTCH. Papaya, Papao, Papay

Carica papaya L.

There are 45 species of Papaya. Indigenous to the West Indies and Central America, bearing fruit when one year old, it is one of the fastest growing trees. Deeply lobed leaves cluster at the top of a hollow trunk under which develop the creamy white fragrant flowers to produce the most popular tropical fruit in the world. Some trees produce fruit bearing flowers, some only pollen bearing flowers, and some both. Most parts of the tree contain papain, a digestive enzyme used extensively as a meat tenderizer, which is tapped from the green fruits and dried. Ceylon and the West Indies supply most of this papain.

PERFUME TREE, Ylang-ylang, Ilang-ilang

SP. Ilán-ilán, Ilang-ilang, Cadmia, Cananga
FR. Ylang-ylang

Cananga odorata syn. *Canangium odorat*

This East Indian 40′ evergreen ornamental is famed as a Perfume Tree, as it produces a valuable oil, known as oil of ilang-ilang. It is distilled from the strongly fragrant droopy 3″ to 5″ yellow flowers, and made into extract as a base for perfume. The tree is in flower and fruit (note berries behind flower above) most of the year. The wood is not durable and will break in a storm. Tree pictured is from Hope Gardens in Jamaica. Also found in S. Florida, Puerto Rico, Virgin Islands, Botanical Gardens in Dominica, St. Vincent, Trinidad, and other Caribbean Islands, but just recently introduced into Central and South America.

POTATO TREE

SP. Arbol de Papa, Azulillo

Solanum macranthum

Remarkable because it is one of the few plants of the Potato Family that takes the form of a tree. The Potato Tree is a beautiful flowering tree from South America that was not associated with the Potato Family until the 19th century. A seed was first cultivated in 1916 in Trinidad from Botanical Gardens in Egypt. The tree is fast growing to 40′ in 2 to 4 years, but is short lived. A tree in Trinidad grew to 30′ with a 30′ spread in two years. Not only do the multi-colored 3″ flowers of violet, pink, and white make a striking display, but the 10″ to 15″ dark green leaves also make a handsome contribution of their own. Tree also has many orange berries as big as golf balls. Blooms the year around. Tree pictured is from San Salvador. Can be seen in Botanical Gardens in Trinidad, Dominica, on other Caribbean Islands, and South America.

POUI, Yellow Poui, Gold Tree, Sunshine Tree, Spring Bells (SEE COVER)

SP. Primavera, El Araguaney, Acapro, Pui Flor Amarillo
FR. Bois d'Evilasse, Guerapariba
DUTCH. Wassiba, Grienhatti, Woite, Bethabara

Cybistax donnell-smithii (formerly *Tabebuia*)

National flower of Venezuela. Known as the largest and strongest tree. A single cubic foot may weigh from 60 to 80 lbs. Few trees can surpass it for durability or beauty in bloom. The rains usually can be expected after it flowers, before the leaves appear.

PINK POUI, Pink Cedar, Pink Trumpet Tree, White Cedar

SP. Roble Blanco, Roble de Yugo, Amapola, Amapa Rosa, Matilisquate
FR. Poirier, Poirier du Pays, Poirier Rouge
DUTCH. Kibra Hacha

Tabebuia pentaphylla
syn. *T. heterophylla*

A cousin to the larger Yellow Poui noted for its valuable timber with wood like oak. It is the national flower of El Salvador. Faster growing Bignonia than the Yellow Poui. Both Pouis can be found through the Caribbean, Bermuda, S. Florida, Puerto Rico, South America and Mexico. Used extensively for ornamentals, and shade trees for coffee and cacao bushes.

ROSE APPLE, Plum Rose, Jambu

SP. Pomarrosa, Pomo, Mazana Rosa, Jambeiro.
FR. Pomme Rose, Pommier Rose, Jambosier
DUTCH. Appelroos, Pommeroos

Eugenia jambos L.

Thought to be the tree representing the theory of creation that bore the golden fruit of immortality. Native of India and Malaya, it is widely cultivated in Tropical America. Planted also in Florida and Bermuda, this 30′ evergreen is a relative of the Malay Apple (see page 45). It has good looking narrow pointed leaves, large yellowish-white showy pompon flowers, and egg shaped edible fruits which have a rose-water taste and odor that persists even in jelly. The wood is used as fuel. It is dull brown, hard and heavy. The tree flowers and fruits throughout the year.

ROSE OF VENEZUELA, Mt. Rose, Large Headed Brownea

SP. Rosa de Monte, Palo de Cruz, Rosa de Cruz, Rosa de Montana

Brownea grandiceps Jacq.

The species name *grandiceps* means large headed. The 8″ flower heads that bloom almost continuously on this 40′ native of Venezuela are lovely. The long drooping leaves protect the large delicate flowers among them from the bright rays of the tropical sun during the day, but at night, the leaves lift and expose the blossoms to the dew. Found in Botanical Gardens in Jamaica (Castleton), Trinidad, and St. Vincent.

SAMAN, Rain Tree, Cow Tamarind, Monkeypod

SP. Samán, Guango, Dormilón, Algarrobo, Cenícero, Carreto, Zorra, Lara, Urero, Genizaro
FR. Arbre à Pluie, Gouannegoul, Samana
DUTCH. Regenboom

Samanea saman

Early travelers reported that the Saman produced rain at night. Legend has it that it "rains from the branches the juice of the cicadas." It is noticeable that the grass is green beneath the tree, no doubt due to the fern-like leaflets closing up at night and the rain falling through. This lofty patriarchal canopied tree, one of the most symmetrical in the world, is covered with delicate tiny pink-tipped tufts from spring to fall. Long black pods dangle from fall to winter. These are relished for fodder. The wood is durable and takes a beautiful finish. Used for furniture—all of the furniture of the airport on Dominica is said to be made from just one Saman Tree.

SANDBOX, Hura, Hura Wood, Possomwood

SP. Javilla, Habillo, Habilla, Jabillo, Tronador, Nune, Ceiba, Molinillo
FR. Sablier, Arbre au Diable
DUTCH. Zandkokerboom, Possentrie
Hura crepitans L.

A full branched deciduous spiny tree from the West Indies, Central, South America, and Mexico with poisonous milky sap which could cause blindness. It has ovate, heart-shaped leaves, a dark maroon flower, and fruits like tiny 3″ pumpkins (see close up above, also right below). These seed capsules are the curious part of the tree as each contains 15 to 20 divisions with round flat crescent seeds. They explode when ripe, throwing these seeds all over. They can be used for jewelry (see below), or the entire fruit can be used as a container for sand for blotting letters and paper weights. Early use in this way gave the tree the name Sandbox (Sablier).

SAUSAGE TREE

SP. Arbol de la Salchicha, Aplo de Salchichón

Kigelia pinnata (Jacq.) D.C.
syn. *K. africana* (Lam.) Benth.

Looking like someone's April Fool joke, this lofty, wide-spreading 50′ tree is grown mainly for its curious, odd sausage-shaped fruits. Weighing 10 to 14 lbs. each, they dangle from the branches for many months. Large, wine colored flowers that bloom at night and have a rather unpleasant odor hang down on cord-like strings and drop each morning after flowering (see close up above). Then the funny grey sausages develop. They are not edible, but are used externally for medicine in Africa. Found in Caribbean Islands, Puerto Rico, Cuba, Gardens in Trinidad, St. Vincent, Dominica.

TAMARIND

SP. Tamarindo, Tamarindade
FR. Tamarin, Tamarinier, Tamarindier
DUTCH. Tamarinde, Tamarijn
Tamarindus indica L.

A well known medium sized 40′ stately tree with a dense rounded crown of feathery leaves. Pale yellow flowers with dark red flower buds are followed by 1½″ x 5″ grey pods containing dark brown edible, though very sour, pulp. Used for a beverage, candy and preserves. Naturalized throughout the tropical world. Legume Family.

WILD TAMARIND, Wild Popinac, Cow Tamarind, Leadtree, West Indies Mimosa, Shack Shack, Acacia

SP. Tamarindillo, Hediondilla, Acacia Palida, Lino, Grandino, Guaje, Uaxim
FR. Barba de León, Grains de Lin Pays, Bois Lolo, Monval
DUTCH. Tumbarabu, Tantan
Leucaena glauca

The small trees that grow along the roadsides of all tropical areas are a common weed of the Legume Family, but they have many uses. The seeds, after softening in boiling water, are strung as jewelry, belts, and other novelties (see picture). Pods and leaves make a high protein fodder for cattle and goats, but horses, donkeys, and hogs lose their hair if they partake of them.

TRAVELLER'S TREE, Traveller's Palm

SP. Arbol del Viajero, Palma del Viajero
Ravenala madagascariensis J. F. Gmel.

A thirsty traveller can always find a drink in a Traveller's Tree. The large hollow leaf-stalks form receptacles which store a quart of water. Although the tree resembles a palm, it belongs to the Banana Family from Madagascar. Its flowers, leaves, and fruit are similar to its cousin, the White Bird of Paradise. The orange "Birds" and Heliconia also belong to this family. These can all be seen pictured in "Tropical Blossoms of the Caribbean" by *Hargreaves.*

Lithographed in U.S.A.
W. A. Krueger Co. Micro-Color®
Brookfield, Wis. U.S.A.